Water-Train

The Most Energy-Efficient
Inland Water Transportation

Synthesis Lectures on Ocean Systems Engineering

Editor
Nikolaos I. Xiros, *University of New Orleans*

The Ocean Systems Engineering Series publishes state-of-the-art research and applications oriented short books in the related and interdependent areas of design, construction, maintenance and operation of marine vessels and structures as well as ocean and oceanic engineering. The series contains monographs and textbooks focusing on all different theoretical and applied aspects of naval architecture, marine engineering, ship building and shipping as well as sub-fields of ocean engineering and oceanographic instrumentation research.

Water-Train: The Most Energy-Efficient Inland Water Transportation
Kurian George
2020

Feedback Linearization of Dynamical Systems with Modulated States for Harnessing Water Wave Power
Nikolaos I. Xiros
2020

Marine Environmental Characterization
C. Reid Nichols and Kaustubha Raghukumar
2020

Water-Train: The Most Energy-Efficient Inland Water Transportation
Kurian George

ISBN: 978-3-031-01364-5   paperback
ISBN: 978-3-031-02492-4   ebook
ISBN: 978-3-031-00320-2   hardcover

DOI 10.1007/978-3-031-02492-4

A Publication in the Springer series
SYNTHESIS LECTURES ON OCEAN SYSTEMS ENGINEERING

Lecture #3
Series Editor: Nikolaos I. Xiros, *University of New Orleans*
Series ISSN
Synthesis Lectures on Ocean Systems Engineering
Print 2692-4420   Electronic 2692-4471

Water-Train

The Most Energy-Efficient
Inland Water Transportation

Kurian George
Retired Executive Engineer at the Kerala State Electricity Board

SYNTHESIS LECTURES ON OCEAN SYSTEMS ENGINEERING
#3

ABSTRACT

In a government-aided research project carried out at Cochin University, the inventor of the Water-Train demonstrated that his invention requires only 24 BTU/ton-km of energy whereas barges use 328 BTU in the same Inland water transportation situation. The use of this Water-Train can invariably curtail, to a large extent, the emission of greenhouse gases thereby decreasing the effect on global warming. Conventional water vehicles use screw propellers which have high reacting energy loss in propulsion whereas the Water-Train relies on the earth for reaction which is an infinite mass causing no reacting energy loss at all. The propelled water takes away a large quantity of kinetic energy ($1/2mv^2$ where its mass is m and velocity is v). Water-Train requires a monorail rigidly fixed to the earth through cross arms and pillars for applying the traction/propulsion force. The reacting body is the earth and so the traction efficiency tends toward 100%. It utilizes low friction of water and also the vehicles are connected serially like a locomotive and hence the wave making and skin resistances are also reduced. The NITIE study conducted earlier in India showed that diesel and electric trains use 166.3 BTU and 105.76 BTU, respectively, for the same purpose.

KEYWORDS

inland water transport, water train, tractor wheel propulsion, propulsion efficiency, energy efficiency, rail connected traction, BTU/ton-km

Contents

Series Editor Foreword xi

Acknowledgments xv

Nomenclature and Abbreviations xvii

1   Introduction 1
   1.1   The Great Need for Energy-Saving Devices 1
   1.2   The Concept 2
   1.3   Water-Train Projects Successfully Carried Out 3
   1.4   Main Advantages of the Water-Train 5
   1.5   Patent 6

2   Energy Efficiency of Conventional Propulsion Systems 7
   2.1   Types of Conventional Propellers 7
   2.2   Elements of the Conventional Propulsion System 7
   2.3   Powers and Efficiencies of the Propulsion System 7
   2.4   Design Control Over the Efficiency of a Conventional
         Propulsion System 11
   2.5   Axial Momentum Theory of the Screw Propeller 11

3   The Concept of the Water-Train 17
   3.1   Starting from the Newton's Third Law of Motion and
         Study of the Energy Shared 17
   3.2   The Driving System of the Water-Train 20
   3.3   The Trailer Bogies 22

4   Fabrication and Testing of the Model in the Second Project 23
   4.1   The Vessels 23
   4.2   The Driving System 23

4.3 The Monorail Track 24

4.4 The Final Set-Up 25

4.5 Results of Model Tests 26

    4.5.1 Model Particulars 26

    4.5.2 Meter Readings 27

    4.5.3 Average Speed, Froude Number, and Reynolds Number of Model 28

    4.5.4 Average Input Power to the System Which is Output Power of Motor 28

    4.5.5 Method of Prediction of Effective Power of 500 Ton Water-Train from the Test Result 30

    4.5.6 Calculation of Energy Efficiency of Water-Train Based on the Model Test 31

4.6 The Model Speed Selected Corresponding to the Full Size Having a Speed of 9.5 km/hr 32

4.7 Ralli Wolf Limited Motor Test Results 35

4.8 Propulsion Efficiency of Freight Transport in India 35

5 Conclusion and Future Work **39**

5.1 Conclusion 39

5.2 Future Work 40

5.3 The Proposed Pilot Project 40

    5.3.1 Bogie/Vessel Details 40

    5.3.2 The Power System 42

    5.3.3 The Flexible Driving Arm 42

    5.3.4 The Suspension System 43

    5.3.5 How the Vessels Move Parallel to the Monorail When the Track is Curved 46

    5.3.6 The Flexible Guiding Arm 46

    5.3.7 The Flexible Couplings 46

    5.3.8 Monorail Track and Crossing Facility for Other Vehicles 47

    5.3.9 Supporting Concrete Pillars 54

5.4 Pilot Project for Commercial Purpose 56

Author's Biography 57

Series Editor Foreword

Water-Train is a new concept intended to transport both passengers and cargo through inland waterways especially in the equatorial region where sea levels are almost steady. The vessels of the Water-Train move one behind another like a locomotive-powered train on land. In the case of Water-Train, a monorail track is used—front vessel or any middle vessel in the train can be the tractor. The front vessel and the last vessel are designed with a hydrodynamic or stream-lined shape to improve efficiency and decrease power requirements. The middle vessels have a rectangular box-like shape. All the vessels are connected to each other using flexible couplings. The vessels are connected to a monorail track supported by concrete pillars. The pillars are fixed in the bed of the waterway at regular intervals and have cross arms at the top to carry the monorail of slender cross-section. The rail is fixed in the cross arm in the inverted position as the vessel's weight is not resting on the rail. The purpose of the rail is for creating the traction and stopping through brakes and all the more for guiding the vessels in the specific route during wind and waves which tend to deviate the vessels from the fixed path parallel to the monorail. The physical basis for the Water-Train is that the rate at which water mass is pushed backward during propulsion is much less than the mass of the vehicle as well as that of the forward-moving water due to wave making and skin friction. Furthermore, since the tandem motion of several vehicles causes the system to reduce the overall wave-making and skin-resistance components faced by the system in its forward motion, the energy required is less.

At the same time, the U.S. military is exploring the viability of unmanned surface vessels, or USVs, to conduct a range of dangerous cross-ocean missions without a crew on board. But while the smaller, versatile watercraft can be useful for tasks involving surveillance, logistics, electronic and expeditionary warfare, and offensive operations, their size, shape, and other components have proven at times to limit the vessels' ability to endure choppy waves.

According to a recent announcement unveiled in late 2019, the Pentagon's research arm aims to improve the long-range operational capabilities of the Navy and Marine Corps' USVs by creating "sea trains" of four or more physically connected vessels, or vessels that are not connected but sail in coordinated formations. The ultimate goal is to develop systems of smart, crewless warships that can travel thousands of ocean miles and perform their own duties, all while exploiting wave-making resistance reductions. Specifically, the Defense Advanced Research Projects Agency (DARPA) Sea Trains program seeks to revolutionize the next generation of unmanned surface vessels, as officials support in the announcement.

Through the program, DARPA envisions the development of a four-vessel-or-more system that can persist through arduous transits of about 7,500 miles, then disaggregate and conduct their own tasks. Alternatively, the vessels may conduct independent yet collaborative operations consisting of transits, loiters, and sprints in varied sea state conditions for more than 1,100 miles. The vessel will then reassemble for transit as the connected Sea Train for another 7,500-mile journey—all without human intervention. The agency is open to any technical approaches that proposers wish to offer but highlighted three that could potentially achieve the overall concept. Officials recommend fleets of connected or connector-less Sea Trains that essentially create a mid-body for the vessel to decrease wave-making resistance while also allowing for the vessels to separate and conduct tactical missions independently. The agency also suggests "formation sea trains," or a fleet made up of four or more vessels that travel closely and exploit wave interference between one another while in transit.

DARPA is very interested in nontraditional approaches that reduce risk early in the program. Early, aggressive, and insightful modeling, analysis, and testing is encouraged and desired.

The program is anticipated to envelop two phases. The first phase will encompass conceptual design, analysis, simulation, scaled model testing, and the second will include updates and additional testing.

Research will also have to address two technical areas over the course of the program. For the first area, researchers will need to develop a conceptual design of an integrated system comprised of the vessels' hull form, connectors, propulsion, and gap mitigation techniques. For the second technical area, participants will be expected to create a dynamic "open-standard, autonomous" control architecture that can monitor environmental conditions the vessels endure in the middle of the ocean, their alignment and spacing, and control solutions to maximize Sea Train efficiency and seaway survivability.

Many of these issues and obstacles for the Sea Train have been success-fully tackled in a proven fashion by the Water-Train concepts. Specifically, three projects were carried out to validate the Water-Train concept and have it ver-ified in the field. The first project was carried out in the Kerala State Science and Technology Museum, Thiruvananthapuram; it was a mini working model outfitted with eight bogies. The energy efficiency of the Water-Train was in-vestigated in the one-tractor, one-trailer system developed in the second project carried out at the Cochin University of Science and Technology. All tests and operations in the first two field-testing projects were carried out in calm water and some experts expressed doubts about how the Water-Train would respond to waves. In effect, a third project was carried out at the Indian Institute of Technology, Kharagpur. Flexible couplings, flexible driving arm, flexible guid-ing arms, and flexible transmission were developed for a larger-scale Water-Train outfitted three bogies. The Water-Train was run in the towing tank on waves created by the wave maker. The test results of this third project proved that the Water-Train performed really well on waves.

So as the Sea Train program seeks to enable extended transoceanic transit and long-range naval operations by exploiting the efficiencies of a system of connected vessels, the Water-Train concept seeks to do so in inland waters and exploiting the benefits of both propulsion over land and movement with low friction when moving through water. Indeed, land vehicles like trucks and trains are relatively more efficient because a major portion of the energy output of the engine is absorbed by the vehicle itself in the form of kinetic energy as the reacting mass is the earth which is practically infinite in value. However, watercraft have the advantage of low friction loss at low speeds as compared to friction between solids. In effect, the Water-Train is the only system of water transportation which eliminates propellers and uses the rail-connected traction taking advantage of low water friction.

The Sea Train program can take the Water-Train concept to the next logical step of development. If successful, it will demonstrate long-range de-ployment capabilities for a distributed fleet of tactical USVs. DARPA will de-velop and demonstrate approaches to overcome the range limitations inherent to medium unmanned surface vessels (MUSVs) by exploiting wave-making resis-tance reductions. DARPA envisions sea trains formed by physically connecting vessels with various degrees of freedom between the vessels, or vessels sailing in collaborative formations at various distances between the vessels, just like the Water-Train concept has pioneered and envisioned as well.

Nikolaos I. Xiros, DEng
Professor, University of New Orleans
Naval Architecture & Marine Engineering
New Orleans, July, 2020

Acknowledgments

I am extremely thankful to the Director of The Kerala State Science and Technology Museum and the State Government for funding the design, fabrication, and installation of a mini working model of the Water-Train in the museum. Also, the KSEB granted me duty leave for five months to complete the above work for which I am very grateful.

I also wish to acknowledge The Kerala State Science, Technology, and Environment Committee for providing financial support for the second project. I am grateful to the Cochin University Authorities, especially the then Vice Chancellor, T.N. Jayachandran, for the various encouragements rendered for the execution of the project. The authorities of KSEB and Kerala Government granted one extra year of duty leave to me during the period in which the major portion of the work in this second project was completed and I thankfully cherish this support. I also wish to acknowledge the authorities of the workshop in the physics department, USIC, and the ship technology department of Cochin University for providing all the facilities available during the fabrication of the model. The various technical advices received from Dr. K. Sathianandan (guide of the project), Dr. Dileep K. Krishnan (Reader) and Manu Korulla (Student) of the Cochin University are thankfully acknowledged. I wish to thank Ralli Wolf Ltd., Bombay for designing, fabricating, and donating the traction motor for the project. The various assistances received from the authorities of the KERI, Peechi is thankfully acknowledged and the roles of Prabhakaran, Ambujakshan, and Sivaraman are worth mentioning.

This useful invention had remained ignored in India. My repeated attempts could not open the eyes of the government officials who could not realize the fact that India is the ideal place for implementing such a highly energy-efficient water transport as the country is near to the equator with many steady waterways. Dr. Nikolas Xiros (Professor of Naval Architecture and Marine Engineering, University of New Orleans, USA) found out the significance of the invention and came forward, making use of his earnest effort for its implementation and so his open mind and wisdom are very gratefully acknowledged.

Kurian George
July 2020

Nomenclature and Abbreviations

B	Breadth of vessel
BTU	British thermal unit
C_{TL}	Thrust loading coefficient
CUSAT	Cochin University of Science & Technology
Fn	Froude Number
IIT	Indian Institute of Technology
KERI	Kerala Engineering Research Institute
KSEB	Kerala State Electricity Board
L_{OA}	Length Overall of vessel
n	Revolutions per second of the propeller
NITIE	National Institute of Industrial Engineering
OPC	Overall Propulsive Coefficient
P_B	Brake Power
P_D	Delivered Power
P_E	Effective Power
Ps	Density of Water
P_T	Thrust Power
Q	Propeller torque
QPC	Quasi-Propulsive Coefficient
Rn	Reynolds Number
R_{TOTAL}	Total Resistance
t	Thrust deduction fraction
T	Propeller thrust
T	Draft of vessel
USIC	University Science Instrumentation Centre
V	Speed of model or prototype
V	Voltage
V_A	Velocity of advance of the propeller
w	Wake fraction

η_B	Propeller Efficiency
η_H	Hull Efficiency
η_I	Ideal Efficiency of Propeller
η_O	Open Water Efficiency
η_p	Propulsion Efficiency
η_R	Relative Rotative Efficiency
η_T	Transmission Efficiency
λ	Scale factor
Δs	Displacement of full-scale prototype (Tons)
Δm	Displacement of Model (Tons)
ss	Wetted surface area of full-scale Water-Train
vs	Speed of the full-scale Water-Train
η	Tractorsystem efficiency
m	Mass
v	Velocity
KE	Kinetic Energy

CHAPTER 1

Introduction

1.1 THE GREAT NEED FOR ENERGY-SAVING DEVICES

Emission of greenhouse gases and the consequent effects on global warming is currently a great problem that humanity is facing. One way to reduce the emission of greenhouse gases is to increase the efficiency of propulsion/traction systems of marine transport. Screw propellers are the most common device used for propulsion of marine vehicles. The best efficiency of a screw propeller in ideal conditions is just about 70%. In shallow water and inland waterways, due to the limited draft of the vessels, the efficiency can be as low as 50%. But it is the **propeller efficiency** and not the **propulsion efficiency**. Any efficiency term is a fraction with the useful part in the numerator as output and in the denominator is the output plus losses which is the input. Here in the term *propeller efficiency*, the axial (parallel to the axis of the propeller) flow of propelled water in the backward direction is the useful part. When its mass decreases the velocity has to increase which is found out by equating the momentums. But the axial flow of propelled water can be treated only as a loss since it doesn't contribute to energy increase of the vessel or overcoming the resistance during the movement of the vessel in the forward direction. So the calculations show that the propulsion efficiency is less than 15% especially in shallow water as 85% of the input energy to the propeller is wasted in the propelled water in the form of kinetic energy. For reducing the kinetic energy gained by the propelled water, the reactive energy loss has to be minimized there by propulsion efficiency can be increased to more than 90%. If earth is used as the reacting body instead of water, then the mass of the reacting body is tending to infinity and so its backward velocity is tending to zero. So the energy loss in the reacting body is zero and the traction/propulsion efficiency is tending to 100% theoretically.

1.2   THE CONCEPT

If one examines the energy transfer mechanism of the water vessels propelled by conventional propellers, it can be seen that there is considerable energy loss in the propeller itself. In the case of water vehicles like boats, barges, and ships the mass of water that is propelled backward per second is much less than the mass of the vehicle and mass of water energized due to wave making and skin resistance resulting in high reactive energy loss. The land vehicles like trucks and trains are relatively more efficient because a major portion of the energy output of the engine is absorbed by the vehicle itself in the form of kinetic energy as the reacting mass is the earth which can be treated as an infinite mass. However, water vehicles have the advantage of low friction loss at low speeds as compared to friction between solids. The energy efficiency data of the Indian Transport Policy Committee would demonstrate this (Table 4.11). Energy requirement per ton kilometer of a diesel train is 166.3 BTU whereas the same for a slow-speed diesel barge is 328 BTU. It may also be noted that the water resistance that opposes the forward motion of the vessel contains two major parts, viz. wave-making resistance and skin friction. Barge trains already in use in Europe and other parts of the world have demonstrated that if the vessels are connected in series like a train and closely packed, the overall wave making and skin resistance will be reduced. One of the methods to improve the efficiency of a water transport system is to replace the conventional propeller system by rail-connected traction, the propulsion/traction is accomplished by using the reactive force from the static monorail rigidly fixed on the ground or to the earth.

In the following sections, the design, development, manufacture, and testing of an entirely new propulsion system is described where the thrust is obtained using tractor wheels running along a monorail which is in the entire length of the route. In this system, the propulsion/traction efficiency can be as high as 90% or even higher compared to a maximum of about 15% of conventional propulsion systems. It is not the propeller efficiency of 70% or 50%, as mentioned earlier, which is wrongly conceived as propulsion efficiency. Propulsion system is not used to drive a Water-Train. It is similar to a barge train with specific advantages like low value of resistance/ton-km and flexibility of size of train.

Water-Train is a new technical concept intended to transport both passengers and cargo through inland waterways especially in the equatorial region where sea level is almost steady. The vessels of the Water-Train move one behind another like a locomotive-driven land train along a monorail track. The front vessel or any middle vessel can be the tractor. The front vessel and the last vessel have hydrodynamic or streamlined shape. The middle vessels have a rectangular box-like shape. All the vessels are connected to each other using flexible couplings. The vessels are connected to a monorail track which is about 3 meters above water level and is supported on concrete pillars fixed in the bed of the waterways at a regular interval of about 5 meters. The pillars have cross arms at the top to carry the monorail (single rail) of slender cross section. The rail is fixed in the cross arm in the inverted position as the vessel's weight is not resting on the rail. The purpose of the rail is for creating the traction and stopping through brakes and all the more for guiding the vessels in the specific route during wind and waves which tend to deviate the vessels from the fixed path parallel to the monorail.

1.3   WATER-TRAIN PROJECTS SUCCESSFULLY CARRIED OUT

Three government-aided projects were carried out in the field of Water-Train.

The first project carried out in the Kerala State Science and Technology Museum, Thiruvananthapuram was a mini working model having eight bogies, as shown in Figures 1.1 and 1.2.

The energy efficiency of the Water-Train was investigated in the one-tractor-one trailer-system developed in the second project carried out at the Cochin University of Science and Technology. The system was operated and tested at KERI, Peechi. The energy efficiency test conducted showed that Water-Train requires only 24 BTU/ton-km where as diesel barges require 328 BTU/ton-km for transportation. Here an electric motor was used as the prime mover. Its energy requirements were found as 24 BTU/ton-km. But if the electric motor is replaced by a diesel engine the proportionate increase of energy is predicted as 37.7 BTU/ton-km ($24 \times \frac{166.3}{105.76} = 37.7$ BTU) which is 11.5% of the energy requirement of diesel barges. The energy efficiency test was supervised by an expert team led by the Professor Dr. Walter Stovhase from Wilhelm Pieck University of Rostock, Germany.

Figure 1.1: Water-Train front view.

Figure 1.2: Water-Train side view.

All the above tests and operations were carried out in calm water and some experts expressed doubts about how the Water-Train will respond to waves. So, a third project was carried out at the Indian Institute of Technology, Kharagpur. Flexible couplings, flexible driving arm, flexible guiding arms, and flexible transmission were developed for a larger-scale Water-Train having three bogies. The Water-Train was made to run in the towing tank on waves created by the wave maker. The test results of this third project proved that the Water-Train performed really well on waves.

1.4 MAIN ADVANTAGES OF THE WATER-TRAIN

- High traction/propulsion efficiency of above 90% compared to that of less than 15% for conventional propulsion in shallow water. Even though the propeller efficiency is 50%, the propulsion efficiency is less than 15% as mentioned earlier.

- Conventional screw propellers operating behind the ship increase the pressure resistance of the ship. The rail-connected traction mechanism used for propulsion in the Water-Train does not cause any increase in vessel resistance.

- Unlike conventional propellers operating in inland water, the propulsion system in the Water-Train located above the water line is not affected by water weeds and floating debris.

- The efficiency of a conventional screw propeller decreases as the thrust-loading coefficient increases. The efficiency of the Water-Train propulsion system is not affected by an increase in thrust requirement.

- The number of vessels can be increased or decreased depending on cargo or passenger availability as in the cases of locomotives.

- The train pattern of motion, vessels one behind the other and closely packed, reduces the overall wave making and skin resistances compared with that of the vessels moving independently. It provides further energy saving over and above the high traction efficiency.

- Ordinary water vehicles, especially speed boats, cause severe damage to the banks where Water-Train is friendly to the shore or banks.

- Since the supporting pillars, monorail, etc., are fixed along the middle line of the waterway and at a convenient height of 3 meters, it does not create any impediment to the operation of other conventional vehicles and their use of the two banks and also for the crossing of barges having large width and height. Figures 5.10 to 5.12 provide the details. Also since the supporting pillars are 50 meters apart, criss-cross movements for the boats are permitted in the entire stretch.

- Water-Train vessels will not sink in water since they are connected with the monorail and so there is better safety.

- Water-Train can share the waterway with the conventional water vehicles like boats and barges as in the case of trams sharing the road with road vehicles like cars and trucks without causing mutual impediments.

- All the more, the great energy-saving advantage of Water-Train will decrease the global warming by reducing greenhouse gas emissions to a considerable extent.

1.5   PATENT

"Water-Train," India Patent 160426, issued on August 27, 1984.

C H A P T E R 2

Energy Efficiency of Conventional Propulsion Systems

2.1   TYPES OF CONVENTIONAL PROPELLERS

1. Screw Propeller

    - Open propeller
    - Ducted propeller
    - Azimuthing propeller

2. Water Jet Propeller

2.2   ELEMENTS OF THE CONVENTIONAL PROPULSION SYSTEM

- Main Engine

- Gear Box

- Propeller Shaft

- Propeller

- Hull

2.3   POWERS AND EFFICIENCIES OF THE PROPULSION SYSTEM

EFFECTIVE POWER P_E

The effective power is the power expended by the hull in moving through water at a given speed V, overcoming the total resistance R_T. Effective power is given by

$$P_E = R_T V.$$

THRUST POWER P_T

Thrust power is the output power of the propeller and is given by

$$P_T = T V_A,$$

where T is the propeller thrust and V_A is the velocity of advance of the propeller, measured relative to surrounding water.

The difference between the ship speed V and the propeller velocity of advance is called wake. The wake fraction is given by

$$w = \frac{V - V_A}{V}.$$

DELIVERED POWER P_D

The power delivered to the propeller by the shaft is called Delivered Power and is given by

$$P_D = 2\pi n Q,$$

where n is the revolutions per second of the propeller and Q is the propeller torque.

BRAKE POWER P_B

Brake power is the output power of the main engine and is the input power to the transmission consisting of gear and shaft.

EFFICIENCY OF THE PROPULSION SYSTEM

Overall Propulsive Efficiency

The overall efficiency of the propulsion system is called Overall Propulsive Co-efficient (OPC) and is given by

$$OPC = \frac{P_E}{P_B}.$$

The efficiency of individual elements of the propulsion system are as follows.

Transmission Efficiency

The transmission consists of the gear and shaft. The transmission efficiency is given by

$$\eta_T = \frac{P_D}{P_B}.$$

The transmission efficiency is generally in the range 95–98%. Losses in the transmission are friction losses in the gear and shaft bearings, and are dependent on the number and quality of shaft bearings and the shaft alignment.

Propeller Efficiency

The input to the propeller is the delivered power and the output from the propeller is the thrust power. Propeller efficiency in the behind condition (propeller working behind the hull) is given by

$$\eta_B = \frac{P_T}{P_D}.$$

The efficiency of a propeller is normally determined by open water tests in a towing tank or cavitation tunnel where a scaled down geometrical similar model propeller is tested without the hull. The efficiency determined from open water tests is called the Open Water Efficiency η_O which is somewhat different from the actual efficiency η_B in the behind condition.

The ratio of propeller efficiency in the behind condition to the propeller efficiency in the open water condition is called Relative Rotative Efficiency η_R.

The propeller efficiency in the behind condition is then given by

$$\eta_B = \eta_O \eta_R.$$

The propeller efficiency in the behind condition can vary widely depending on the design constraints and the operating conditions. Even in ideal conditions, the best efficiency that can be expected from a properly designed propeller is between 65% and 70%.

Hull Efficiency: Hull-Propeller Interaction

There is a two-way interaction between a screw propeller and the hull. The hull modifies the flow coming into the propeller, causing the phenomenon called wake:

$$wake = V - V_A$$

and the wake fraction is given by

$$w = \frac{V - V_A}{V}.$$

The propeller working behind the hull modifies the flow coming around the hull in the stern region by accelerating it in the sternward direction and also imparting a rotational velocity to the flow. The increased flow velocity in the stern region due to the working of the propeller causes a decrease in pressure there, leading to an increase in pressure resistance. The increase in resistance of the ship due to propeller working is alternatively considered as a decrease in the thrust produced by the propeller. The main part of the thrust overcomes the towed resistance of the ship while the remaining part (thrust deduction) is used to overcome the additional resistance caused by the rotation of the propeller. The thrust deduction is given by

$$T - R_T,$$

where R_T is the total resistance of the ship in the towed condition with speed V without propeller and T is the thrust generated by the propeller working behind the ship to move the ship at the same speed V.

The thrust deduction fraction is given by

$$t = \frac{T - R_T}{T}.$$

Hull efficiency is the ratio of power actually expended by the hull in overcoming the towed resistance R_T to the thrust power and is given by

$$\eta_H = \frac{P_E}{P_T} = \frac{R_T V}{T V_A} = \frac{1 - t}{1 - w}.$$

2.4   DESIGN CONTROL OVER THE EFFICIENCY OF A CONVENTIONAL PROPULSION SYSTEM

Hull Efficiency

The hull efficiency is generally slightly greater than unity, since wake fraction is generally greater than the thrust deduction fraction. The propeller designer has practically no control over the hull efficiency which is a function of wake fraction and thrust deduction fraction. The hull form is optimized to minimize the water resistance and the propeller is generally optimized for the greatest efficiency. The hull efficiency is then automatically fixed by the hull-propeller combination.

Transmission Efficiency

Transmission efficiency is generally high in the range of about 0.95 to 0.98. The designer can aim for smaller losses in the shaft bearings by having a shorter shaft. However, a marginal increase in transmission efficiency is not going to have a great effect on the OPC.

Propeller Efficiency

This is the critical efficiency that every designer tries to maximize, by carrying out a rigorous optimization exercise. Given a set of design constraints, there is an upper limit to which the efficiency can be increased. As the thrust loading on the propeller increases, the efficiency decreases, as shown in the following section. In the most ideal conditions, the upper threshold value is just about 0.75, which means at least 25% of the power is lost in the conventional propeller.

2.5   AXIAL MOMENTUM THEORY OF THE SCREW PROPELLER

The axial momentum theory tries to explain how a screw propeller generates thrust. The theory uses momentum and kinetic energy concepts to derive mathematical expressions for the thrust and ideal efficiency of the propeller.

ASSUMPTIONS

1. The propeller is idealized as a circular disk of diameter D, same as the diameter of the propeller. The disk accelerates the flow coming around the hull in the sternward direction.

2. The flow is frictionless and the inflow into the propeller disk is unlimited.

   The cylindrical column of water passing through the disk is called the race column. Upstream of the disk at section 1, just where the flow begins to accelerate, the race column has an average axial velocity V_A at the disk, i.e., section 2, the axial velocity increases to $V_A(1 + a)$. Downstream of the disk at section 3, the flow achieves the maximum axial velocity $V_A(1 + b)$. Continuity equation demands that the race column will contract between sections 1 and 3 due to increase in velocity, with a diameter equal to the disk diameter at section 2. a and b are called axial inflow factors. Obviously, a and b are positive and $b > a$.

MASS FLOW RATE THROUGH DISK

Disk area is $A_o = \pi \frac{D^2}{4}$.

Volume flow rate, $Q = $ Disk area $\times$ Axial Velocity at disk $= A_o V_A(1 + a)$.

Mass flow rate $= \rho Q = \rho A_o V_A(1 + a)$ where ρ is the density of water.

PROPELLER THRUST

The disk accelerates the flow passing through it, increasing its momentum. The rate of change of momentum of the flow must equal the force exerted by the disk on the flow as per Newton's second law. As per Newton's third law, the flow exerts an equal and opposite reaction on the disk, thus creating the propeller thrust T:

$$T = \textit{Rate of change of momentum of race column}$$

$$T = \rho Q \, (V_3 - V_1) = \rho Q \, [V_a \, (1 + b) - V_A] = \rho Q V_A b.$$

Plug in the expression for mass flow rate to get

$$T = \rho A_o V_A^2(1 + a)b.$$

WORK DONE BY THE DISK AND KINETIC ENERGY OF THE RACE COLUMN

The work done by the disk in accelerating the race column must be equal to the increase in kinetic energy of the race column.

- Work done by the disk in unit time = Thrust × Axial Velocity at the disk = $T V_A (1 + a)$.

- Plug in the expression for T to get:

  Work done by disk in unit time = $\rho Q V_A^2 b (1 + a)$.

- Increase in kinetic energy in unit time $= \frac{1}{2} \rho Q \left(V_3^2 - V_1^2 \right) = \frac{1}{2} \rho Q \left[\{V_A (1 + b)\}^2 - V_A^2 \right] = \rho Q V_A^2 b \left(1 + \frac{b}{2} \right)$.

- Equating work done and increase in kinetic energy,

$$ a = \frac{b}{2}. $$

  The above equation shows that half the increase in flow velocity is achieved by the time the flow reaches the disk.

IDEAL EFFICIENCY

Ideal efficiency of the propeller is defined as the ratio of useful work done by the disk to the total work done by the disk. It can also be defined as

$$ \frac{\textit{Work done by fluid reaction on the disk}}{\textit{Work done by disk on fluid passing through it}}. $$

The ideal efficiency is then

$$ \eta_I = \frac{T V_A}{T V_A (1 + a)} = \frac{1}{(1 + a)}. $$

THRUST LOADING COEFFICIENT

Thrust loading coefficient C_{TL} is a non-dimensional coefficient defined as follows:

$$ C_{TL} = \frac{T}{\frac{1}{2} \rho A_O V_A^2}. $$

Plug in the expression for T in the above equation to get

$$C_{TL} = \frac{\rho A_o V_A^2(1 + a)b}{\frac{1}{2}\rho A_O V_A^2} = 2(1 + a)b.$$

Since $b = 2a$, the above equation becomes

$$C_{TL} = 4a(1 + a).$$

RELATION BETWEEN IDEAL EFFICIENCY AND THRUST LOADING COEFFICIENT

$$\eta_I = \frac{1}{(1 + a)}.$$

Solving for a from above, $a = \frac{1}{\eta_I} - 1$. Plug into the equation for C_{TL} to get

$$C_{TL} = 4\left(\frac{1}{\eta_I} - 1\right)\frac{1}{\eta_I}.$$

Put $\frac{1}{\eta_I} = \alpha$ and plug into above equation to get a quadratic equation in terms of α.

$$4\alpha^2 - 4\alpha - C_{TL} = 0.$$

Solve for α. The second root with the negative sign before the discriminant is discarded as this leads to a negative value for efficiency.

$$\alpha = \frac{1 + \sqrt{1 + C_{TL}}}{2}.$$

Hence, the ideal efficiency is given by

$$\eta_I = \frac{2}{1 + \sqrt{1 + C_{TL}}}.$$

The above equation gives some valuable insight into the efficiency of the propeller. As the thrust loading coefficient increases, efficiency of the propeller decreases. Hence, it is a good idea to try to reduce the thrust loading coefficient. Since

$$C_{TL} = \frac{T}{\frac{1}{2}\rho A_O V_A^2}.$$

For a given required thrust, C_{TL} can be reduced by

- increasing the diameter of the propeller leading to an increase in A_o,

- having multiple screws where the total thrust can be divided between two or more propellers, thus reducing the thrust of each propeller.

It is obvious that for a dramatic increase in propeller efficiency, an entirely new and radically different propeller must be designed, as outlined in the following chapters.

CHAPTER 3

The Concept of the Water-Train

3.1 STARTING FROM THE NEWTON'S THIRD LAW OF MOTION AND STUDY OF THE ENERGY SHARED

Newton's third law says "to every action there is an equal and opposite reaction." Here, action and reaction are forces. Then it can be inferred that for applying a force there must be at least two bodies. A single force cannot exist on its own. But force can exist only in pairs.

Let us imagine two bodies in space which are originally at rest and mutually in contact with each other, as shown in Figure 3.1. X Y is a separating imaginary line and A and B are the two bodies.

The mass of A and B are equal and denoted by "m". If we want to accelerate A to a velocity "v" from rest, then energy gained by A is $1/2mv^2$. But A alone cannot be energized to velocity "v". Then B also will be correspondingly energized to velocity "v", as shown in Figure 3.2. Then energy gained by B is also $1/2mv^2$. The percentage efficiency of propulsion/traction of $A = \frac{\frac{1}{2}mv^2 \times 100}{\frac{1}{2}mv^2 + \frac{1}{2}mv^2} = 50\%$. This 50% efficiency exists when the two masses of A and B are equal.

Then let us imagine the two masses are not equal, B is half that of A. Mass of A is "m" and mass of B is "$1/2m$", as shown in Figure 3.3. But momentum of the two bodies must be equal which are "$m \times v$" in both directions. Then velocity attained by the lighter body (B) is $2\,v$. Then kinetic energy gained by A is $1/2mv^2$ and that by B is $1/2\left[1/2m \times (2\,v)^2\right] = 1/2\left[1/2m \times 4v^2\right] = 1/2\left[2mv^2\right] = mv^2$. Then percentage efficiency of propulsion/traction of $A = \frac{\frac{1}{2}mv^2 \times 100}{\frac{1}{2}mv^2 + mv^2} = 33.3\%$.

Similarly it can be shown as in Figure 3.4 that if the mass of B is $1/4$ "m", then % efficiency of propulsion/traction of $A = 20\%$.

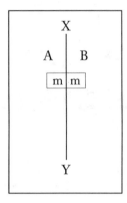

Figure 3.1: Two bodies in space at rest.

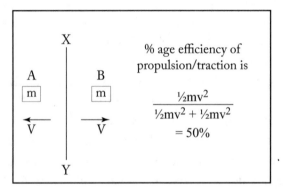

Figure 3.2: Efficiency of propulsion/traction when a force is applied.

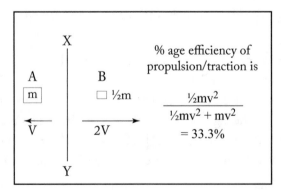

Figure 3.3: Efficiency of propulsion/traction when mass of *B* becomes 1/2 m.

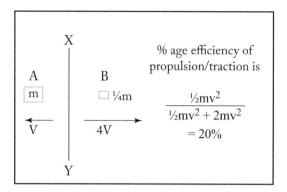

Figure 3.4: Efficiency of propulsion/traction when mass of B becomes 1/4 m.

When the mass of the reacting body B is reduced below that of A as shown, then the propulsion/traction efficiency of A is found going below 50%. But if the mass of the reacting body is increased above that of A, it is found going above 50%. If the mass of B can be made infinitely big as earth then the reacting energy loss is tending to zero and the propulsion/traction efficiency is found theoretically tending to 100%.

According to Newton's third law of motion "to every action there is an equal and opposite reaction." This has really carried us to a wrong notion that the energy gained by the body A due to the influence of the force action is also equal to the energy gained by the body B due to the influence of the force reaction. This can happen only when the two masses A and B are equal. When the masses are different, energy shared is not equal. **Energy gained by a body due to the influence of a force is inversely proportional to its mass. Lighter bodies gain more energy and heavier bodies gain less energy.** It is a universal law to be accepted just like Newton's third law of motion. The best example is the gun firing a bullet. If it is assumed that the mass of the bullet is $\frac{1}{100}$ of the mass of the gun, then the velocity of the bullet will be 100 times that of the gun. It is derived from equating the momentums, i.e., $m_1 v_1 = m_2 v_2$ where m_1 and v_1 are the mass and velocity of the bullet and m_2 and v_2 are the mass and velocity of the gun. From the above equation of momentum, it can be derived that, $v_1 = \frac{m2v2}{m1} = \frac{100m1v2}{m1} = 100v_2$. Here v_2 is the final velocity of the gun after firing. Then kinetic energy gained by the bullet is 100 times that of the gun. Keeping this important principle in mind, if the propulsion of a water vehicle is analyzed, it will be seen that the mass of propelled water per second is much less than the mass of the vehicle plus mass of the affected water in wave making and skin resistances. Then naturally the axial velocity in the backward direction of the propelled water is very high and the kinetic energy gained also is very large giving a very high reactive energy loss. But if earth is used as the reacting body and it is done through the monorail and supporting pillars fixed on earth and since Earth's mass is infinitely big, then traction/propulsion efficiency is tending to 100%. This is the main idea behind the rail-connected traction. Since no weight of the driving and guiding arms is resting on the rail and the same are borne by the vessel as the suspension system is used, the structural cost of the rail track and also the friction between the driving wheel and rail is very much reduced and so the traction efficiency comes up to more than 90% in actual case.

Also, since the vessels are aligned one behind the other and closely packed like in a locomotive, the overall wave making and skin resistances are reduced which result in very high traction efficiency of 37.7 BTU when the barges use 328 BTU for transporting one ton of weight to a distance of 1 kilometer.

3.2   THE DRIVING SYSTEM OF THE WATER-TRAIN

In driving a vehicle of this type, the driving force from the engine is transmitted through gears and shafts to the driving wheel, which operates in a horizontal plane by touching the side of the monorail which is fitted above the water level. The monorail alignment along the side of the train and the layout are shown in Figure 3.5.

The tractor bogie has two arms on the side of the monorail, one fitted in one end and other in the opposite end which is the driving arm. The driving wheel is fixed on this arm and this wheel comes in contact with the side of the monorail while in operation. Its position is in between the bogie and the monorail. A pressure wheel fitted to the driving arm operates on the other side of the rail. This wheel is also set in the horizontal plane. The pressure exerted by the pressure wheel to the side of the rail can be controlled through a spring mechanically or by hydraulic pressure or by air pressure from a compressed air chamber. On the application of thrust to the pressure wheel, it moves on to the side of the rail and exerts a force. The reaction on this wheel now causes the driving wheel to move toward the rail. Thus, the rail will be pressed in between

Figure 3.5: One tractor and one trailer system used in KERI, Peechi for energy efficiency test.

the driving wheel and the pressure wheel. In the present design gaps of the order of a few millimeters prevail between the two wheels and the sides of the rail. In the pressed position, on rotating the prime mover, the driving wheel will roll along the side of the rail without slipping. There are two more wheels on this arm and these wheels are expected to roll along the top and bottom surfaces of the rail on occasions when the train is subjected to rolling and pitching due to wind or waves. However, during normal conditions of the water surface there will be sufficient gaps between these two wheels and the rail.

The guiding arm is used only for guiding purposes. This arm contains four wheels all in fixed positions. These wheels have sufficient clearance with the rail and they come in contact with the rail only when external forces disturb the smooth motion of the vehicle. Another important feature relating to the design of the arms is their ability to adjust with variations of the water level. A more elaborate self-adjusting flexible device has also been developed independently and could be adopted to cater to such situations. It is elaborately illustrated in the future development work in Chapter 5.

3.3   THE TRAILER BOGIES

Any number of trailer bogies can be connected to the tractor just like an ordinary locomotive, subject to the maximum power output of the prime mover. The energy efficiency of the system is expected to be higher when the numbers of trailer bogies are more and when they are closely packed. To compensate for unequal loading and the consequent unequal levels of the bogies, appropriate coupling devices are required in the successive bogies. A flexible coupling has been developed for connecting the bogies. The trailer bogie used here is provided with a guiding arm fitted in the rear portion. For the model developed and tested at KERI, Peechi, India, one tractor bogie and one trailer bogie were used. The driving system was also considerably simpler. As the water level was maintained steadily and both the bogies were loaded equally, rigid coupling and rigid driving and guiding arms were used.

CHAPTER 4

Fabrication and Testing of the Model in the Second Project

4.1 THE VESSELS

The vessels of the model prototype were fabricated at the workshops in the departments of Physics and Ship technology of the Cochin University of Science and Technology. A one tractor–one trailer system was fabricated. The physical size of these two vessels was identical (length 6.14 meters, breadth 0.55 meter and depth 0.45 meter giving a full load draft of 0.30 meter), as shown in Table 4.1. One end of each bogie was given a streamlined shape. The other ends of these two bogies were blunt and flat in shape and these ends were connected together (permitting only a small gap between them to avoid eddies during motion) using a rigid connector or coupling. The streamlined shape ends were at the two ends pointing in opposite directions. The gross weight or mass displacement of the system when loaded with two people (driver and passenger), equipment, and a suitable quantity of ballast (sand bags) was 1.5 tons.

4.2 THE DRIVING SYSTEM

The prime mover is a universal motor (AC/DC) which can develop the required starting torque. The rotor speed of the motor is 2800 rpm and shaft speed (output speed) is 175 rpm which gives a gear ratio 1/16. The speed of the motor is varied by changing the input A.C. supply using a dimmerstat. At 200 Volt A.C., the required output speed of 175 rpm is obtained. The direction of rotation of the motor can be reversed by changing the current supply to the field winding by providing forward-reverse switch in series with the field winding of the motor. The mechanical output power of the motor is only 44 Watts during testing on the Water-Train. The efficiency of the motor is low due to its small size and also

Table 4.1: Main particulars of model

Length overall, L_{OA} (m)	12.18 m
Length on water line, L_{WL} (m)	12 m
Breadth (m)	0.55 m
Draft (m)	0.3 m
Mass displacement, Δ (t)	1.5 tons
Wetted surface area, S (m^2)	11.064 m^2
Speed (m/s)	0.96

due to the use of worm gears for the speed reduction. This special motor was designed and fabricated by M/s. Ralli Wolf Ltd. Bombay and was supplied as a gift to the project. The manufacturer had conducted the energy efficiency test in the motor and is found around 25% (Tables 4.9 and 4.10). The motor is fitted on the driving arm of the tractor and the driving wheel is coupled to the output shaft of the motor. In the final set up, the output shaft is vertical and the driving wheel rotates in a horizontal plane. The input voltage and current are read from the voltmeter and ammeter mounted in front of the driver and observer. In addition, a sensitive energy meter (with 3000 disc rotations per kWH and of 2.5A/5A rating single phase) is also connected along with the two meters. The speed of the vehicle is determined by noting the time required to cross a marked distance of 50 meters at steady speed.

4.3   THE MONORAIL TRACK

The crucial and sensitive part of the project is the setting of the monorail track. Since there is no suitable waterway available in the CUSAT campus, I had to depend on external sources. The cost of model tests in towing tanks at IIT Madras or IIT Kharagpur was beyond the limited budget of the project. Being a delicate power testing experiment, rivers or busy backwater shores were not considered suitable for the purpose. Fortunately, the special water tank (water current meter rating tank) built by the KERI, Peechi certainly was the best available for this purpose in Kerala. This tank is 100 meters long, 2.4 meters wide and 1.75 meters deep. The water level of the tank can be controlled by adjusting the inlet valve. On normal days one can obtain still water at a constant level in the tank.

Figure 4.1: **Dr. Walter Stovhase** and the author on the Water-Train during the test.

The monorail track was made using 100 x 50 mm size M.S. joist. The monorail system is fixed on 40 numbers, of specially made M.S. supports (using 40 x 40 x 6 mm size M.S. angles) with bolts and nuts. The cantilever supports are fixed outside the retaining wall using concrete and rubbles.

The monorail track is fitted slightly away from the middle line of the tank so that the Water-Train can be operated along the middle line of the tank. The track is 100 meters long of which the first 25 meters is left for starting and accelerating while the last 25 meters is used for decelerating and stopping the vehicle. The remaining 50 meters along the middle portion is the steady speed region and this region is marked by two pegs.

4.4   THE FINAL SET-UP

The M.S. joists for the monorail and the M.S. angles for the supports fabricated in the CUSAT campus were transported to the site at KERI, Peechi water tank and the most precision work of laying the monorail track on cantilever supports was carried out with much care. The track was made absolutely straight and horizontal through repeated corrections and erections. Also, the required single-phase A.C. supply was drawn from the overhead electric lines passing above the tank using one post, standard fuse-switch system and wires. About 60 meters of long loose wire was used to give supply to the moving Water-Train from the middle point of the tank.

After completing the monorail track work, the two bogies fabricated in the CUSAT campus were also transported to the site at KERI, Peechi. They were set on the water and appropriate sand loading was made so that both the bogies float at the same level. The driving arm and the two guiding arms were fixed to the bogies. The motor-gear unit (Prime mover) was fixed in position in the driving arm. The driving wheel, pressure wheel, and their various operations were properly aligned. Electrical connections to the dimmerstat and the various meters, switches and motor were given. Several test runs were necessary to optimize the conditions. The trial runs have demonstrated that the train reached a uniform speed during the first 25 meters and the motion of the vehicle during the next 50 meters distance could be carefully studied for the power consumption.

4.5   RESULTS OF MODEL TESTS

The details of the test arrangements and the results are given in the following sections. The tests were witnessed by a team from CUSAT consisting of the following members:

1. Dr. Walter Stovhase, Visiting Professor, Wilhelm–Pieck University, Rostock, Germany.

2. Dr. K. Babu Joseph, Professor and Head, Department of Physics, Cochin University of Science and Technology, Kerala, India.

3. Dr. Dileep Krishnan, Reader, Department of Ship Technology, Cochin University of Science and Technology, Kerala, India.

4. Dr. V. P. Narayanan Nampoothiri, Professor, Department of Physics, Cochin University of Science and Technology, Kerala, India.

4.5.1   MODEL PARTICULARS

The main particulars of the model are given in Table 4.1.

Table 4.2: Time and speed runs

Run Number	Direction	Time Taken (s)	Speed (m/s)
1	Forward	50.0	1.00
2	Reverse	53.0	0.94
3	Forward	50.7	0.99
4	Reverse	54.0	0.93
	Average	51.9	0.96

Table 4.3: Energy consumption during runs

Run Number	Direction	Energy Consumption (Wh)
1	Forward	11.50
2	Reverse	11.50
3	Forward	11.25
4	Reverse	10.75

4.5.2   METER READINGS

Stopwatch Readings

Time required to travel 50 meters at steady speed and the speed achieved during each run is given in Table 4.2.

Voltmeter Readings

The voltmeter reading was constant at 200 V for all runs, both forward and reverse.

Energy Meter Reading

The energy consumption in Wh during each run of 50 m distance at steady speed is shown in Table 4.3.

Ammeter Reading

The ammeter reading in amps during each run of 50 m distance at steady speed is shown in Table 4.4.

When the no-load current input of the motor was measured at 200 V using an ammeter the readings were 1.2 A (Reverse) and 1.5 A (Forward). The

Table 4.4: Ammeter readings during runs

Run Number	Direction	Ammeter Reading (A)
1	Forward	1.70
2	Reverse	1.40
3	Forward	1.67
4	Reverse	1.44

Table 4.5: Froude Number and Reynolds Number of the model

Average speed from Table 4.2 (m/s)	0.96
Coefficient of Kinematic Viscosity, ν	0.892×10^{-6}
Froude Number, F_n	0.0888
Reynolds Number, R_n	12966980

corresponding current values at 200 V supply furnished by M/s. Ralli Wolf Limited as test results (see Tables 4.9 and 4.10) is 1.45 A (Reverse) and 1.6 A (Forward). This may be due to the effect of the gears and other moving parts in contact with each other getting smoother by continuous working and also due to the difference in the accuracy of the two ammeters. So a correction factor of 0.25 A in the reverse direction and 0.1 A in the forward direction were required to be added to the ammeter readings obtained in the test.

4.5.3   AVERAGE SPEED, FROUDE NUMBER, AND REYNOLDS NUMBER OF MODEL

Froude Number and Reynolds Number of the model are given in Table 4.5 (note that in Table 4.5 $\nu = 0.892 \times 10^{-6}$ for fresh water at 24°C).

4.5.4   AVERAGE INPUT POWER TO THE SYSTEM WHICH IS OUTPUT POWER OF MOTOR

See Table 4.6.

Table 4.6: Input power, efficiency, and output power of motor

Direction	Voltmeter Reading (V)	Ammeter Reading (A)	Input Current after Correction (A)	Power Input to Motor (W) (Table 4.9, Table 4.10)	Efficiency of Motor (Table 4.9, Table 4.10)	Power Output of Motor (W)
Forward	200	1.7	1.8	335	12.9	43.22
Reverse	200	1.42	1.67	300	14.84	44.52
					Average	43.87

4.5.5   METHOD OF PREDICTION OF EFFECTIVE POWER OF 500 TON WATER-TRAIN FROM THE TEST RESULT

As the losses in the traction system are unknown, effective power of the model P_{EM} is calculated for 3 different value of losses viz. 1%, 5%, and 10%.

P_{EM} is then scaled to 500 t displacement full-size Water-Train using Froude's Law as follows.

- Scale Factor $\lambda = \sqrt[3]{\frac{\Delta S}{\Delta M}}$,

  where

  - ΔS = Displacement of full-scale prototype Water-Train [Tons].
  - ΔM = Displacement of model [Tons].

- Total resistance of model, $R_{TM} = \frac{P_{EM}}{V_M}$,

  where

  - V_M = Velocity of model in m/sec.
  - P_{EM} = Effective power in kW.

- Total resistance coefficient of model, $C_{TM} = \frac{R_{TM}}{\frac{1}{2} P_M S_M V_M^2}$,

  where

  - P_M = Density of water in which model was tested.
  - S_M = Wetted surface area of model.

- Frictional resistance coefficient of model C_{FM} is calculated using I.T.T.C. Friction Line.

  - $C_F = \frac{0.075}{\left[Log_{10} R_n - 2\right]^2}$,

    where R_n = Reynold's number.

- The Residual Resistance coefficient of Model $C_{RM} = C_{TM} - C_{FM}$.

- As the Froude Number is kept constant for model and full-scale Water-Train, i.e., $\frac{V_M}{\sqrt{g\, L_M}} = \frac{V_S}{\sqrt{g\, L_S}}$.

- According to Froude's Law, $C_{RS} = C_{RM}$,

  where

  - C_{RS} = Residuary resistance coefficient of the full-scale Water-Train.

  - C_{RM} = Residuary resistance coefficient of the Model.

- The frictional resistance coefficient of full-scale Water-Train C_{FS} is calculated using the I.T.T.C. friction line. The total resistance co-efficient of full-scale Water-Train is then given by, $C_{TS} = C_{RS} + C_{FS} + C_A$, where

  - C_A is a roughness allowance whose value is assumed as 0.0004.

- Total resistance of full-scale Water-Train $R_{TS} = C_{TS}\frac{1}{2}P_S S_S V_S^2$, where

  - P_S = Density of water.
  - S_S = Wetted surface area of full-scale Water-Train = $S_M \times \lambda^2$.
  - V_S = Speed of full-scale Water-Train = $V_M \sqrt{\lambda}$.

- Effective Power of full-scale Water-Train is $P_{ES} = R_{TS} V_S$.

- The input power to system is then given by $\frac{P_{ES}}{\eta\eta}$, where

  - η = Tractor system efficiency.

4.5.6   CALCULATION OF ENERGY EFFICIENCY OF WATER-TRAIN BASED ON THE MODEL TEST

System Efficiency of the Model

- Model Length L_M at water line = 12.00 m.

- Model Displacement Δ_M = 1.5 tons.

- Model Speed V_M = 1.00 m/sec.

- Model Froude No. Fn = $\frac{V_M}{\sqrt{g\ L_M}}$ = 0.0922.

- Power input to system P_M = 43.87 Watts.

- E_M = Energy/ton kilometer for model is given by

- Power input × time taken to travel 1 km ÷ Displacement.

- $E_M = \frac{43.87 \; x \; 1}{1000 \; x \; 3.6 \; x \; 1.5} = 8.124 \times 10^{-3}$ KWh/ton-km.

- $= 27.72$ BTU/ton-km.

Here the motor losses have not been taken into account since these losses are much more than in the actual case due to the extremely small size of the model motor which uses worm gears.

Prediction of Energy Efficiency for Full-Scale Prototype "Water-Train"
- Proposed full-scale displacement $= 500$ tons.

- Scale factor $\lambda = \sqrt[3]{\frac{500}{1.5}} = 6.934$.

- $L_S = $ Full-scale length of Water-Train $= 6.934 \times 12 = 83.208$ m.

- Full-scale speed $V_S = V_M \times \sqrt{\frac{L_S}{L_M}} = V_M \times \sqrt{\lambda} = 2.633$ m/s,
  $= 9.479$ Km/h.

Since the power loss in the tractor system could not be measured precisely, the effective power required to push the model at that speed, i.e., P_{EM} could not be calculated which prevented a definite prediction for full-scale. However, it is estimated that the power loss in the tractor system will not exceed 10%. Hence, the energy efficiency for full-scale prototype was calculated for 3 values of tractor system loss; i.e., 1%, 5%, and 10%. See Table 4.7.

4.6 THE MODEL SPEED SELECTED CORRESPONDING TO THE FULL SIZE HAVING A SPEED OF 9.5 KM/HR

Speed of the full-size barge (500 tons) is restricted within 10 km (9.5 km in this case). The normal water depth in the inland waterways is about 2 meters. The model speed corresponding to 9.5 km/hr of the full size is 1 m/s (3.6 km/hr). The extract received from the National Transportation, Planning and Research Center (Table 4.8) shows that when the depth of waterway is below 2.5 meters and draught is less than 2 meters, then the speed has to be restricted to 7 to 9 km/hr for the full size barge. The energy consumption for a big barge in shallow water dramatically goes high and it is not advisable to move such a vehicle at high speeds. In addition to power loss, bank erosion is prohibitively high when speed increases.

Table 4.7: The results of the calculation of energy efficiency of Water-Train

Tractor System Loss	1%	5%	10%
Effective Power of model P_{EM} (KW)	0.0434	0.0417	0.0395
Total resistance of model R_{TM} (KN)	0.0434	0.0417	0.0395
Total resistance coefficient of Model C_{TM}	7.845×10^{-3}	7.538×10^{-3}	7.140×10^{-3}
Frictional resistance coefficient of model C_{FM} (From I.T.T.C. friction line)	2.851×10^{-3}	2.851×10^{-3}	2.851×10^{-3}
Residuary resistance coefficient of model C_{RM} = $C_{TM} - C_{FM}$	4.994×10^{-3}	4.687×10^{-3}	4.289×10^{-3}
Residuary resistance coefficient of prototype C_{RS} = C_{RM}	4.994×10^{-3}	4.687×10^{-3}	4.289×10^{-3}
Frictional resistance coefficient of prototype C_{FS} (From I.T.T.C. friction line)	1.837×10^{-3}	1.837×10^{-3}	1.837×10^{-3}
Total resistance coefficient of prototype C_{TS} = $C_{RS} + C_{FS} + C_A$ ($C_A = 0.0004$)	6.8314×10^{-3}	6.5244×10^{-3}	6.1264×10^{-3}
Total resistance of prototype R_{TS} (KN)	12.597	12.031	11.297
Effective power of prototype P_{ES} (KW)	33.17	31.68	29.74
Input power to tractor stem P_s (KW)	33.50	33.34	33.05
Energy consumption of tractor system in K.W.h./ton Kilometer	7.068×10^{-3}	7.0347×10^{-3}	6.9735×10^{-3}
Energy consumption of tractor system in B.T.U/ton kilometer	24.124	24.009	23.8
$P_s \times \dfrac{1}{9.4788} \times \dfrac{1}{500} \times \mathbf{3413} = \mathbf{BTU/ton\text{-}Kilometer}$			

Table 4.8: The extract received from National Transportation, Planning, and Research Center (*Continues.*)

Sl No.	Type of Vessel	Water Depth (in meters)	Draught (in meters)	Average Power (kW)	Average Speed (km/h)
1	Bulk Carrier B1	5.00	2.80	900	15
		3.50	2.80	650	12
		2.50	2.00	150	8
		2.00	1.50	300	9
2	Bulk Carrier B2	5.00	2.80	900	15
		3.50	2.80	650	12
		2.50	2.00	150	8
		2.00	1.50	300	9
3	Bulk Carrier B3	5.00	2.80	900	15
		3.50	2.80	650	12
		2.50	2.00	150	8
		2.00	1.50	300	9
4	Car Carrier	5.00	1.60	400	14
		3.50	1.60	450	13
		2.50	1.60	150	8
		2.00	1.30	160	7
5	Push Boat PB	5.00	2.50	1300	15
		3.50	2.50	1000	12
		2.50	2.00	300	7
		2.00	1.50	250	7
6	Bulk Carrier B LNG	5.00	2.80	900	15
		3.50	2.80	650	12
		2.50	2.00	150	8
		2.00	1.50	300	9
7	LNG Carrier LNG1	5.00	2.30	900	15
		3.50	2.30	800	13
		2.50	2.30	150	7
8	LNG Carrier LNG2	5.00	2.30	700	14
		3.50	2.30	900	13
		2.50	2.30	150	7

Table 4.8: (*Continued.*) The extract received from National Transportation, Planning, and Research Center

9	Tanker T1	5.00	2.80	900	15
		3.50	2.80	700	12
		2.50	2.00	150	8
		2.00	1.50	300	9
10	Tanker T2	5.00	2.80	900	15
		3.50	2.80	700	12
		2.50	2.00	150	8
		2.00	1.50	300	9
11	Container Vessel CO1	5.00	2.80	900	15
		3.50	2.80	650	12
		2.50	2.00	150	8
		2.00	1.50	300	9
12	Container Vessel CO2	5.00	2.80	900	15
		3.50	2.80	650	12
		2.50	2.00	150	8
		2.00	1.50	300	9
13	RoRo Vessel	5.00	1.60	400	14
		3.50	1.60	450	13
		2.50	1.60	150	8
		2.00	1.30	160	7

4.7   RALLI WOLF LIMITED MOTOR TEST RESULTS

See Tables 4.9 and 4.10.

4.8   PROPULSION EFFICIENCY OF FREIGHT TRANSPORT IN INDIA

See Table 4.11.

Table 4.9: Test results of the special motor designed and fabricated by M/s. Ralli Wolf Limited—Clockwise direction

Direction of rotation : C.W					
Supply voltage : 165 V					
Sr. No.	Input Current Amps.	Input Power Watts	Speed RPM	Torque kg-cm	Efficiency %
1	1.55	230	150	0	0
2	1.6	240	148	10	6.4
3	1.75	250	144	20	11.9
4	1.8	260	139	30	16.68
5	1.95	280	132	40	19.61
6	2.15	305	127	50	21.65
7	2.3	325	121	60	23.23
Supply voltage : 180 V					
1	1.5	230	176	0	0
2	1.65	260	168	10	6.72
3	1.75	280	160	20	11.88
4	1.8	305	154	30	15.75
5	2.05	330	149	40	18.78
6	2.15	345	143	50	21.55
7	2.3	360	138	60	23.92
Supply voltage : 200 V					
1	1.45	260	197	0	0
2	1.5	270	195	10	7.5
3	1.6	290	188	20	13.48
4	1.75	320	180	30	17.55
5	1.85	340	173	40	21.16
6	2.05	365	165	50	23.5
7	2.20	390	159	60	25.44
Supply voltage : 220 V					
1	1.4	280	233	0	0
2	1.5	305	224	10	7.6
3	1.65	330	213	20	13.42
4	1.75	370	199	30	16.78
5	1.85	400	190	40	19.76
6	2.15	430	184	50	22.25
7	2.25	455	182	60	24.96
Note: For 175 RPM at 60 kg-cm, Supply voltage = 210 V, in C.W direction					

Table 4.10: Test results of the special motor designed and fabricated by M/s. Ralli Wolf Limited—Counter-clockwise direction

Direction of rotation : C.C.W					
Supply voltage : 165 V					
Sr. No.	Input Current Amps.	Input Power Watts	Speed RPM	Torque kg-cm	Efficiency %
1	1.6	259	169	0	0
2	1.8	285	164	10	5.9
3	2.0	300	160	20	11.09
4	2.05	315	154	30	15.25
5	2.15	325	149	40	19.07
6	2.3	350	142	50	21.09
7	2.45	365	136	60	23.25
Supply voltage : 180 V					
1	1.6	270	198	0	0
2	1.65	280	192	10	7.14
3	1.75	300	184	20	12.75
4	1.8	320	176	30	17.16
5	2.0	340	169	40	20.67
6	2.2	370	161	50	22.62
7	2.85	390	154	60	24.64
Supply voltage : 200 V					
1	1.6	310	218	0	0
2	1.65	320	214	10	6.9
3	1.8	335	209	20	12.9
4	1.85	360	199	30	17.24
5	2.65	390	192	40	20.48
6	2.10	410	185	50	23.46
7	2.3	430	178	60	25.83
Supply voltage : 220 V					
1	1.75	360	237	0	0
2	1.8	380	231	10	6.3
3	1.9	410	220	20	11.16
4	2.05	440	212	30	15.03
5	2.2	450	203	40	18.76
6	2.3	480	201	50	21.77
7	2.45	510	194	60	23.74
Note: For 175 RPM at 60 kg-cm, Supply voltage = 200 V, in C.C.W direction					

Table 4.11: The extract received from National Transportation, Planning, and Research Center

Propulsion efficiency of freight transport in India		
Mode	Capital Cost (a) (paise/ton-km)	Fuel Efficiency (BTU/ton-km)
Steam Train (b)	5.90	2764.80
Diesel Train (b)	5.90	166.30
Electric Train (b)	6.55	105.76
Diesel Truck (c)	11.23	1587.30
Barge	5.0	328.00
(a) Total capital invested divided into a number of ton-km performed in a year		
(b) Density class 20,000 to 30,000 net-ton per day		
(c) 7.5 ton pay load, 40 km average speed		
Source: NITIE Study as quoted by National Transport Policy Committee, Govt. of India		

CHAPTER 5

Conclusion and Future Work

5.1 CONCLUSION

The experiments carried out at KERI, Peechi, India clearly demonstrates that the propulsive efficiency of the traction wheel propulsion system can exceed 90% with a self-adjusting spring-loaded traction system compared to that of about 50% for a conventional screw propeller (which gives a propulsion efficiency less than 15%) with restricted diameter in shallow water. The energy efficiency of a prototype Water-Train at a speed of 9.6 km/hr in shallow water is about 24 BTU/ton-km compared to that of about 328 BTU/ton-km for a barge with conventional propulsion system operating at a similar speed in shallow water. The Water-Train combines the twin advantages of lower resistance coefficient of a barge train and the higher propulsive efficiency of the traction wheel propulsion system.

Of all the transport vehicles, it has been established that the electric train is the most energy efficient with a per ton-kilometer energy requirement of 105.76 BTU as per the NITIE study reported by the National Transport Policy Committee, Government of India (Table 4.11). The efficiency figure supplied by M/s. Bharat Heavy Electricals Ltd., Bhopal points out that the efficiency of traction motors in the electric trains is about 88%. It is safe to conclude that a diesel Water-Train will consume only below 15% of the fuel consumed by a diesel barge operated at the same speed and gross weight. The present studies thus establish unambiguously that the monorail system of traction using electric power is four times more efficient than the most efficient traction of electric trains. Also, a diesel Water-Train compared to a diesel barge of the same weight and speed would require only below 15% of the fuel requirement of diesel barges. Since the electric Water-Train is found using 24 BTU/ton-km, it can be estimated that a diesel Water-Train will consume

$$\frac{24 \times 166.3}{105.76} = 37.7 \text{ BTU/ton-km},$$

where the per ton-km energy requirement of diesel train is 166.3 BTU and that of electric train is 105.76 BTU, then the ratio of energy requirement of diesel Water-Train and diesel barge is estimated as

$$\frac{37.7}{328} = 0.115 = 11.5\%.$$

5.2   FUTURE WORK

The work on the Water-Train will not be complete without building and running a prototype in an inland waterway. A proposal has been submitted to the government of Kerala for implementation of a pilot project. Initial meetings have been held and the response of the stakeholders has been encouraging.

5.3   THE PROPOSED PILOT PROJECT

A proposal is under the consideration of the government of Kerala to implement a pilot project to run a Water-Train for passenger transport in the Edapally Canal which runs through the heart of the busy metropolis of Kochi, often called the commercial capital of Kerala State. The stretch of the canal from Lulu Mall Junction to Chempumukku with a length of 2.6 km is considered for the project.

5.3.1   BOGIE/VESSEL DETAILS

The pilot project has four bogies/vessels, one tractor, and three trailers. Tractor can be a middle vessel. Each of them can carry about 54 passengers, totally 216 nos. Vessel size is of length 15 m, breadth 3 m, and height 2.5 m. Maximum draft permitted is 0.3 meter, as shown in Figure 5.1. Then maximum displacement is 13.5 metric tons. Unladen weight is 4.5 to 5 tons. When the vessel is loaded with the normal capacity of 54 passengers having an average weight of 75 kg, then the total load of passengers is $54 \times 75 = 4$ tons. Unladen weight is 5 tons giving a total weight in the fully loaded condition as 9 tons. Then the vessel of the train will sink 0.2 meters in water. But rarely the vessel has to be overloaded especially during festival seasons. It may require overloading the vessels

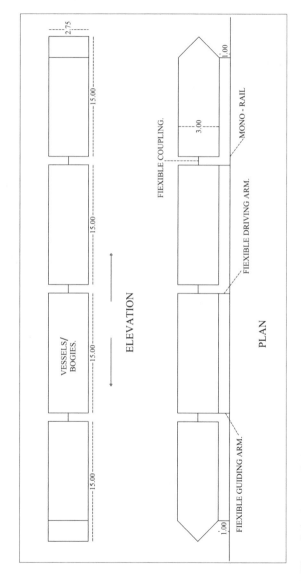

Figure 5.1: Vessel design and size.

to carry even 120 passengers instead of the permitted 54 passengers. Then the depth of draft is around 0.31 meter and the total weight is 14 tons.

5.3.2   THE POWER SYSTEM

The prime-mover is a D.C. motor working at 110 Volt. A DC grid is provided in the entire route and brushes are fitted to draw the power from the grid.

5.3.3   THE FLEXIBLE DRIVING ARM

The most challenging part of the future development is making flexible driving arm, flexible guiding arms, and flexible couplings. The driving arm carries a bracket holding the driving motor, driving wheel, pressure wheel, and the guiding wheels one in the top and the other in the bottom of the rail which touch the rail only occasionally as there is a small gap between the rail and the wheels. The pressure wheel is controlled pneumatically using compressed air or hydraulically. For the best efficiency, the pressure should be more during starting and reduced during running. The rail track is fixed to earth at a height of 3 to 3.5 meters above water level. The driving arm's bracket carrying the driving motor, driving wheel, pressure wheel, and guiding wheels is also steadily moving parallel to the monorail track. But the vessels in the Water-Train undergo, rolling, pitching, and heaving in addition to the level variation caused by tides and changing load condition. The driving arm is expected to act in between the fixed monorail and the vessels which undergo the movements mentioned above.

The driving arm is fitted in the middle vessel/bogie and such Water-Train can move in either direction. The front end and rear end of it is given hydro-dynamic or streamlined shape and that triangular portion of space in the front and back can be used as driver's cabin. Both of such cabins can be used according to the direction in which the Water-Train is moving. When the direction is changed in the return trip, the driver has to shift to the cabin on the other end so as to sit and control the vehicle from the front. The driver can control the Water-Train in both directions sitting in the two cabins. The middle bogie/vessel which acts as the tractor has two arms: one driving and the other guiding. All the remaining bogies have only one guiding arm and no driving arm fitted on them. The only difference between the driving arm and guiding arm is that the latter contains no motor, driving wheel and pressure wheel. In

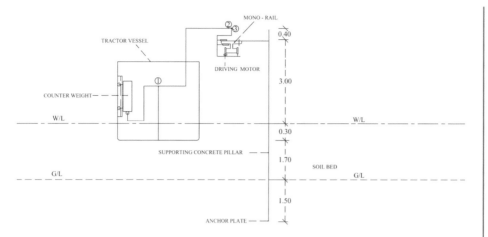

Figure 5.2: Flexible driving mechanism.

place of the driving wheel and pressure wheel, two guiding wheels are provided to make the Water-Train to move parallel to the monorail.

5.3.4   THE SUSPENSION SYSTEM

The bracket carrying the above wheels and driving motor is suspended from the driving arm, as shown in the single line diagram of Figure 5.2. The other end of the arm carries a counter weight below which a wheel is rolling to either side according to the variations found in the arm. Somewhere in the middle of the arm is supported by the vessel base on fulcrum point (1) which is shown in detail in Figure 5.3. In other words, it acts as a common balance, with the weight of the bracket carrying the driving motor, etc., on one side and the counter weight on the other side. The moments of the two weights are almost equal. So the entire weights of the driving arm counter weight, etc. are resting on the base of the vessel. No weight of the same is resting on the monorail. It avoids load on the pillars and foundation and also on the monorail and cross arm. It enables the system to work without vibration and noise. Also, it helps to improve the energy efficiency of traction as it reduces the friction between wheels and monorail.

When the vessel goes down in water due to low tide or increase in load, it can be seen that the distance between the fulcrum point and monorail increases. But the arm length cannot be changed. So there is a tendency for pulling the vessel toward the monorail which increases the load on the monorail and there

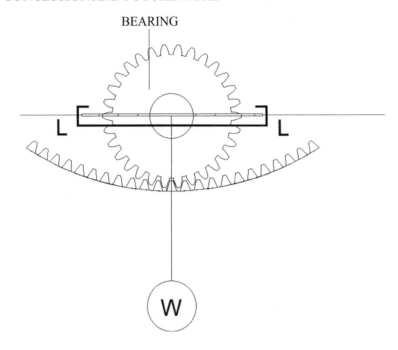

Figure 5.3: Enlarged view of point (1) in Figure 5.2.

is energy loss as the vessel is forced to move toward the rail. Here, since the fulcrum is connected with a gear wheel (rack and pinion), it can roll then on the linear gear on which it is resting. The fulcrum and gear wheel moves toward right keeping the distance steady, as shown in Figure 5.3. When the vessel moves up due to tidal effect and decrease in load, it rolls toward the left again keeping the distance steady. Also, when normalcy is returned, the gear wheel comes and occupies the central point as the linear gear has its lowest point in the center and it goes up toward right and left and it automatically works due to gravity. It maintains the constant distance between the vessel and the monorail even though the vessel moves up and down. Thus, the heaving of the vessel is accommodated between the fixed monorail. In addition, the bearing at point (3) in Figure 5.2 is shown enlarged as in Figure 5.5 gives freedom to the arm to turn up and down without changing the position of suspension.

Then the rolling process of the vessel is analyzed. The fulcrum is in the middle point of the cross section of the vessel. So when the vessel rolls the position of the fulcrum in the middle point will not change as it rolls about

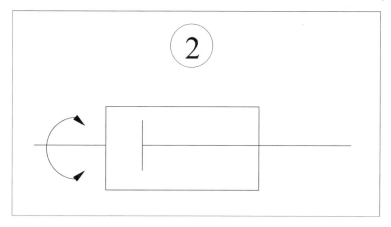

Figure 5.4: Enlarged view of point (2) in Figure 5.2.

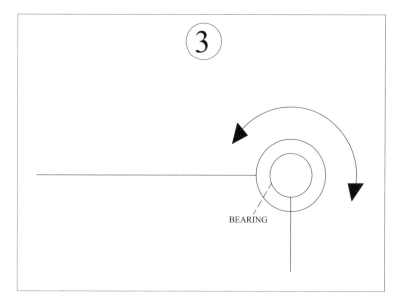

Figure 5.5: Enlarged view of point (3) in Figure 5.2.

the fulcrum (even though the point of fulcrum will go up and down). Relative movement or rotation takes place in the bearing on which the fulcrum operates. Thus, the vessel is rescued from the change made by rolling the vessel.

The bearing at point (2) in Figure 5.2 is shown enlarged, as in Figure 5.4, making the arm free when pitching occurs and no load or force is transmitted to the rail. The wheels that roll along the bottom of the counter weight have a play for 0.2 meter so that when the position of fulcrum moves toward right or left, the wheels will be supporting the counter weight which then goes up or down to correct the alignment.

5.3.5 HOW THE VESSELS MOVE PARALLEL TO THE MONORAIL WHEN THE TRACK IS CURVED

The biggest hurdle in such a mechanism is the possibility of the vessel to move toward or away from the monorail, when the Water-Train moves along a curved track. Then due to centrifugal force, the vessel has the tendency to move away or toward the monorail. But a weight (W) is suspended at the point of fulcrum and it acts as a pendulum of a clock. So when the Water-Train takes a curve parallel to the track toward right, then the weight relatively moves leftward or clockwise and the lock (L) provided goes and engage with the gear wheel arresting its rotation. So that it will not roll. Also, the same process will take place when the vessel takes a curve toward left. But the arm can turn about the fulcrum without arresting its natural movement.

5.3.6 THE FLEXIBLE GUIDING ARM

It contains only four guiding wheels. In place of the driving wheel and pressure wheel, guiding wheels on either side of the rail is provided in the bracket carrying the wheels. It is lighter than the driving arm. The same suspension system can be used in the guiding arms also.

5.3.7 THE FLEXIBLE COUPLINGS

Another important component is flexible couplings to be used between vessels. The tidal effect cannot make the vessels to have relative vertical motion as tides influence all the vessels simultaneously. But unequal loading will make the vessels have individual vertical motion. The flexible coupling has three components made together. A long and smooth pipe having a length of 1.5 meters fitted in the rear side of a vessel (vessels are moving towards left), as shown in Figure 5.6. This pipe passes through the hole of same diameter made in a metal plate piece, as shown in point (1) of Figure 5.6, which is shown enlarged as in Figure 5.7, so that the plate can slide up and down without much play. The other end of the plate is fitted to a hinge like or bush-bearing mechanism which makes a second plate to move horizontally, as shown in point (2) of Figure 5.6 which is shown enlarged as in Figure 5.8. The other end of the second plate is connected

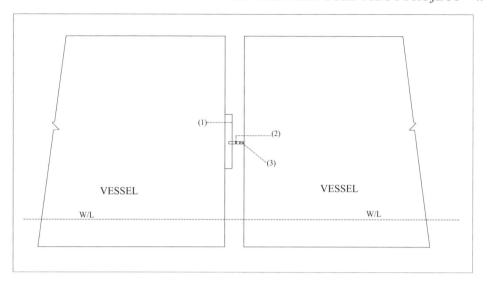

Figure 5.6: Flexible coupling connecting two vessels.

to the bearing of the hinge. Then there is another plate fixed to the second plate through a bearing. This third plate is fixed to the front side of the second vessel, as shown in point (3) of Figure 5.6 which is shown enlarged as in Figure 5.9. Then relative movement of the two vessels in the vertical direction or heaving is permitted by the sliding of the pipe through the hole of the thick plate. Also, the rolling and pitching action of the individual vessels also is permitted through the hinge and bearing provided. All these mechanisms work automatically taking energy from the waves causing pitching, rolling and heaving. The sliding pipe, hinge, and bearing require good lubrication.

5.3.8 MONORAIL TRACK AND CROSSING FACILITY FOR OTHER VEHICLES

The most significant challenge in Water-Train is to provide a monorail track in the waterway/canal without causing hindrance to the movement of ordinary water vehicles like boats and barges. So it is more advisable to avoid the two banks of the canal for fixing the monorail track. The track can be made along the middle of the canal at a good height of 3 meters from the water level during high tide. In equatorial countries, water level variation (not in rivers) in back-waters and sea is below 0.6 meters. The monorail can be fixed on concrete pillars normally at 5 meters apart in the canal bed. Since there is 5 meters wide and 3

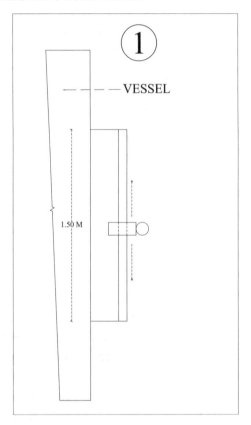

Figure 5.7: Enlarged view of point (1) in Figure 5.6.

meters high opening for crossing the rail track, it is easy for the boats to have criss-cross movement without any difficulty. If crossing of larger width is required, along with the monorail a truss can be fitted even to have a 10 meters wide crossing facility, as shown in Figure 5.10. At certain intervals if barges have to cross the rail track, then cantilever truss using two frameworks and revolving motors are used, as shown in Figure 5.11. When crossing is required the two truss pieces holding the rail will revolve to the vertical position, as shown in Figure 5.12 giving an opening of even 20 meters. When the Water-Train comes, it will revolve to the closed position allowing it to move continuously. The two positions, one "open" and the other "closed" can be displayed through indicator lamps of red and green colors, giving appropriate guidance to both the drivers of Water-Train and the crossing barges. So the introduction of Water-

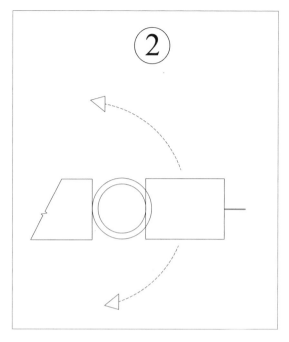

Figure 5.8: Enlarged view of point (2) in Figure 5.6.

Train in the man-made canals and back waters will not create any problem for the movement and crossing for both country boats or power boats or even big barges. From all these it can be seen that Water-Train can have friendly existence with other water vehicles in the same waterway just like trams sharing the road with road vehicles like cars and trucks. No mutual problem is created. The posts or pillars carrying the electrical lines and rail for the trams are not considered as an impediment for the road traffic. Trams are so common in European cities. Here in the case of Water-Train high energy efficiency and mass transportation will become possible. At the time when conventional water vehicles have only propulsion efficiency below 15%, Water-Train has 90% traction efficiency which saves a lot of fuel and energy. Also, the train pattern of vessel arrangement reduces the energy lost in wave making and skin resistance. Thus, it can reduce to a large extent global warming which is really the most relevant problem of the world during this period. In addition, the rail-supported Water-Train has better safety aspects as vessels will not sink in water.

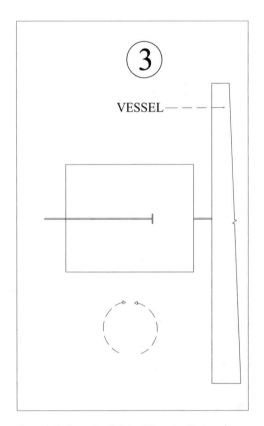

Figure 5.9: Enlarged view of point (3) in Figure 5.6.

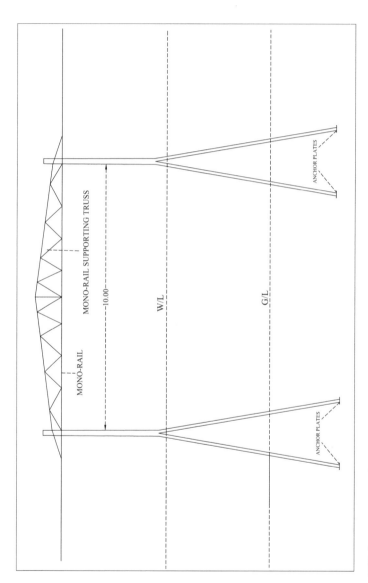

Figure 5.10: Wider crossing device with fixed truss.

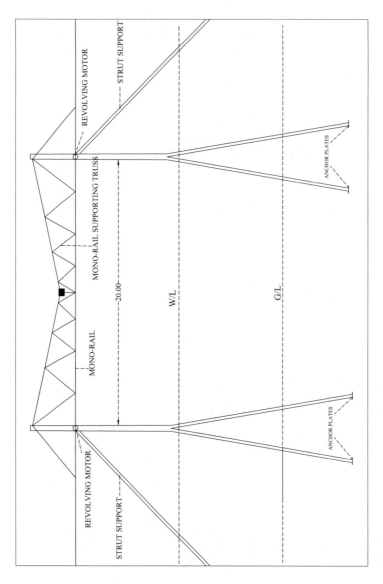

Figure 5.11: Track crossing device in the closed position.

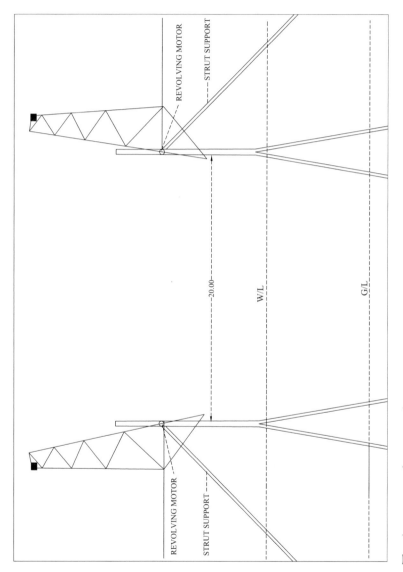

Figure 5.12: Track crossing device in the open position.

5.3.9   SUPPORTING CONCRETE PILLARS

Short concrete pillars (around 7 meters long) will be sufficient for supporting the monorail track. Since no weight of the Water-Train including the driving arm and guiding arm are resting on the slender monorail track, the supports and their foundation need not be very strong which reduces the cost of construction. The foundation for the supports need not be built up from the rocky layer underground. The man made canal has a constant depth of water of about 2 meters. When it crosses backwaters the route can be selected along the shallow region. When rare deep regions come, there stronger foundation becomes unavoidable. Usually 1.5 meters depth can be given to the trench for the pillars, 2.00 meters length, 1.00 meter width will be sufficient for the trenches. The bed of the trench has to be made more loads bearing by applying the internationally accepted sand/chip piling. No cement concrete need to be applied in the trench as slow yielding to load can occur to the pillars which have to be corrected during the maintenance and repair of the track. Therefore, mud concreting is sufficient. 40 mm and 20 mm broken stones have to be mixed together and then again mixed with sand and filled in the trench and rammed using vibrators. The concrete pillars have to be provided with an anchor plate at the bottom with the shape, as shown in Figure 5.13. Cast iron may be the ideal material for making the anchor plates as it is resistant to rust formation and erosion. After completing the sand/chip piling of the trench, a layer of mud concrete having a depth of 0.3 meter to be applied and vibrated for making the bottom surface of the trench more load bearing before placing the anchor plates and the readymade pillars. Then above the anchor plate about 1.2 meters depth of mud concrete can be placed layer by layer and vibrated and thus the pit has to be filled up to the ground level.

The pillars provided in the curved track have to bear higher lateral forces and so an "A" type pillar having more width in the bottom has to be designed in such cases which can bear more lateral load, as shown in the same Figure 5.13. Since sand/chip piling is adopted and no cement is used, the cost for the foundation can be very much reduced. Another option for supporting the track is by fixing the concrete pillars in one of the two banks and fixing long cantilever cross arms in the top so that the vessels will be having sufficient clearance from the bank. This will reduce the construction cost also. The cross-arms fixed at a height of 3 meters above the ground level will not cause obstacle to the move-

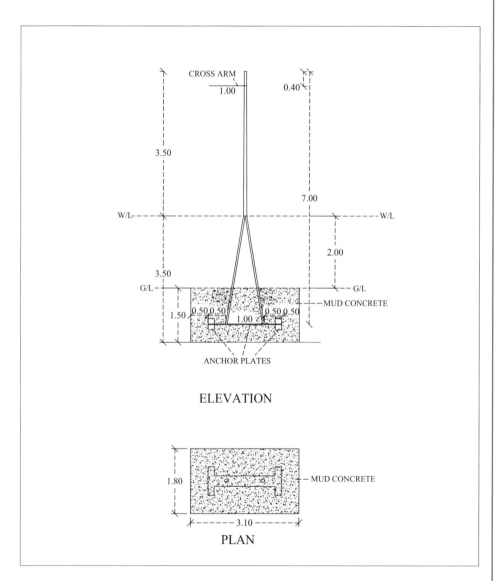

Figure 5.13: Supporting concrete pillar for curved track.

ment of the cargo and passengers in the banks and platform. Also since the monorail track and concrete supports are overhead and having sufficient vertical and horizontal clearance, it will not affect the activities of the boats in the banks on either side. Since the concrete supports are not fixed in the middle of the canal in the second option the entire width of the canal can be used for navigation as the bifurcation of the canal is avoided. The mixed transportation of Water-Train and boats will become very attractive and lucrative later.

The main obstacle caused to boats and barges service is the presence of thick water hyacinth formed in all waterways in India which wind in the propellers and arrest its rotation. But Water-Train has no propeller and so no problem from such water plants and floating debris. It is a great advantage in avoiding propulsion. So both the running cost including fuel and the capital cost can be brought down to a considerable extent in the case of Water-Train which makes it very attractive in financial point of view and environmental aspects.

5.4   PILOT PROJECT FOR COMMERCIAL PURPOSE

After completing the required trials and tests, it is proposed to handover the system to Kochi Metro Ltd. or State Water Transport Department for the transportation of passengers from the LULU MALL Junction to Chempumukku in Kochi, two busy landmarks having tremendous passenger potential.

Author's Biography

KURIAN GEORGE

Kurian George is a respected inventor who has made many contributions to the concept of energy efficiency. He has received many awards for his inventions and holds many patents to his name. He is an Electrical Engineer by profession and had worked for the Kerala State Electricity Board in Kerala, India for most of his professional life until he retired in 1999. Even after retirement he continues to work on developing new ideas solving problems that do not otherwise seem to have a clear solution.

Kurian George first became popular for his invention of the Water-Train, a novel concept that has proved to be the most energy-efficient mode of transportation in the world. In addition, he had made multiple inventions in the area of Electrical Field Engineering like the ergonomic ladder design that is safe and lightweight for use by the field staff. He also redesigned the streetlight reflectors used across the state of Kerala by the electricity board, for which he received multiple awards in honor of his invention.

After retiring from his day job he went on inventing many more practical inventions like better agricultural practices in rubber plantations, water conservation techniques in subterranean rock layers using the Retro-Borewell method, an economic car washing device, an effective way of removing iron from well water without using any chemicals, etc. All of his inventions are rooted in the concept of increasing efficiency, reducing wastage, and improving the quality of life for human kind.

Printed in the United States
by Baker & Taylor Publisher Services